NELSON
EQAO

GRADE 3

Reading
and Writing

NELSON

This workbook belongs to:

_____

NELSON

COPYRIGHT © 2019 by Nelson Education Ltd.

ISBN-13: 978-0-17-688601-1
ISBN-10: 0-17-688601-X

Printed and bound in Canada
1 2 3 4   21 20 19 18

For more information contact Nelson Education Ltd., 1120 Birchmount Road, Toronto, Ontario M1K 5G4. Or you can visit our website at nelson.com.

Credits

Photos
cover: Hybrid Images/Getty Images
6: Nerthuz/Shutterstock.com; 7: freesoulproduction/Shutterstock.com; 8: TAW4/Shutterstock.com; 9: Valeri Hadeev/Shutterstock.com; 16: (mice illustration) Barbara Spurll, (busy street) BABAROGA/Shutterstock.com; 17: (mice illustration) Barbara Spurll, (people crossing street) William Perugini/Shutterstock.com; 18: (mice illustration) Barbara Spurll, (garbage can) Gino Santa Maria/Shutterstock.com; 19: (mice illustration) Barbara Spurll, (mailbox on street) Rolf_52/Shutterstock.com; 20: (mice illustration) Barbara Spurll, (farm with cows) Matthew Jacques/Shutterstock.com; 21: (mice illustration) Barbara Spurll, (subway) Mikael Damkier/Shutterstock.com; 28–30: (illustrations) Matt Roussel; 36: Marko Heuver/Shutterstock.com; 37: Collins93/Shutterstock.com; 38: Ksenia Ragozina/Shutterstock.com; 39: Hibbl/Shutterstock.com; 44: valleyboi63/Shutterstock.com; 45: sianc/Shutterstock.com; 46: valleyboi63/Shutterstock.com; 47: Stolyevych Yuliya/Shutterstock.com; 52: Sergei Bachlakov/Shutterstock.com; 53: (corn) Songsak P/Shutterstock.com, (tools) Snowleopard1/Shutterstock.com; 54: (canoe) Tony Moran/Shutterstock.com, (snowshoeing) Lane V. Erickson/Shutterstock.com; 55: Marilyn Angel Wynn/Nativestock.com/The Canadian Press; 60: Tishchenko Dmitrii/Shutterstock.com; 61: Igor Stramyk/Shutterstock.com; 62: Alexander Sviridov/Shutterstock.com; 63: Daniel Huebner/Shutterstock.com; 68: Sergey Zaykov/Shutterstock.com; 69: schankz/Shutterstock.com; 70: Sergey Zaykov/Shutterstock.com; 71: Sergey Zaykov/Shutterstock.com; 76: schankz/Shutterstock.com; 77: (castings) CHAINFOTO24/Shutterstock.com, (worm illustration) Bart Vallecoccia; 78: Alexander Sviridov/Shutterstock.com; 79: Katya Shut/Shutterstock.com; 84: Frans Lemmens/Photographer's Choice/Getty Images; 85: Regien Paassen/Shutterstock.com; 86: (Umm al-Ma Lake) Patrick Poend/Shutterstock.com, (cupping sand in hands) Dmitry Rukhlenko/Shutterstock.com; 87: Neil Bromhall/Shutterstock.com; 92: Andriy Blokhin/Shutterstock.com; 93: Debu55y/Shutterstock.com; 94: IlkerErgun/Shutterstock.com; 95: Ilona Koeleman/Shutterstock.com; 102: Feel good studio/Shutterstock.com; 103: Jan Danek jdm.foto/Shutterstock.com; 104: Galyna Andrushko/Shutterstock.com; 105: andreiuc88/Shutterstock.com.

Text
6–9: © Nelson Education; 16–21: © Nelson Education; 28–30: © Nelson Education; 36–38: © Nelson Education; 44–47: © Nelson Education; 52–55: © Nelson Education; 60–63: © Nelson Education; 68–71: © Nelson Education; 76–79: © Nelson Education; 84–87: © Nelson Education; 92–95: © Nelson Education; 102–105: © Nelson Education.

Contents

About the EQAO Test

What is EQAO testing?

In Ontario, children in Grades 3, 6, and 9 are required to take a test to assess their literacy skills. The Education Quality and Accountability Office (EQAO) is responsible for this provincial assessment program. The literacy assessments consist of multiple-choice and open-response questions that cover the Ontario Language curriculum for reading and writing.

The *Nelson EQAO Grade 3 Reading and Writing Workbook* is designed to provide students with the opportunity to answer the types of questions they will encounter on the EQAO assessment. This book covers the reading and writing skills your child is learning in school.

At Nelson, we believe in empowering every child with the tools to be successful. This is why we work with respected educators across Canada to develop resources that are aligned to provincial curricula and support the learning journey in school and at home.

How can I help prepare my child for EQAO testing?

The *Nelson Grade 3 EQAO Reading and Writing Workbook* features reading selections and two types of questions: multiple choice and open response. Your child will read a short selection and then answer multiple-choice and open-response questions for that selection. These questions will help your child become more comfortable with the EQAO test format.

Open-response questions require both short and long written answers that assess your child's ability to both read and write. Their responses should demonstrate how well they can communicate their ideas, as well as the content of those ideas. That is, many of the questions require your child to show comprehension of the reading selection. Other questions will use the selection as a springboard, requiring a thoughtful written response to a creative writing prompt.

Sample answers are provided for many of the open-response questions. For other questions, suggested criteria are included for what a good answer will look like.

In addition, to help your child prepare for the EQAO test, you might want to review the tips on page 118 together.

REWARD CONTRACT

When you complete a test in your *Nelson EQAO Grade 3 Reading and Writing Workbook*, colour in a circle.

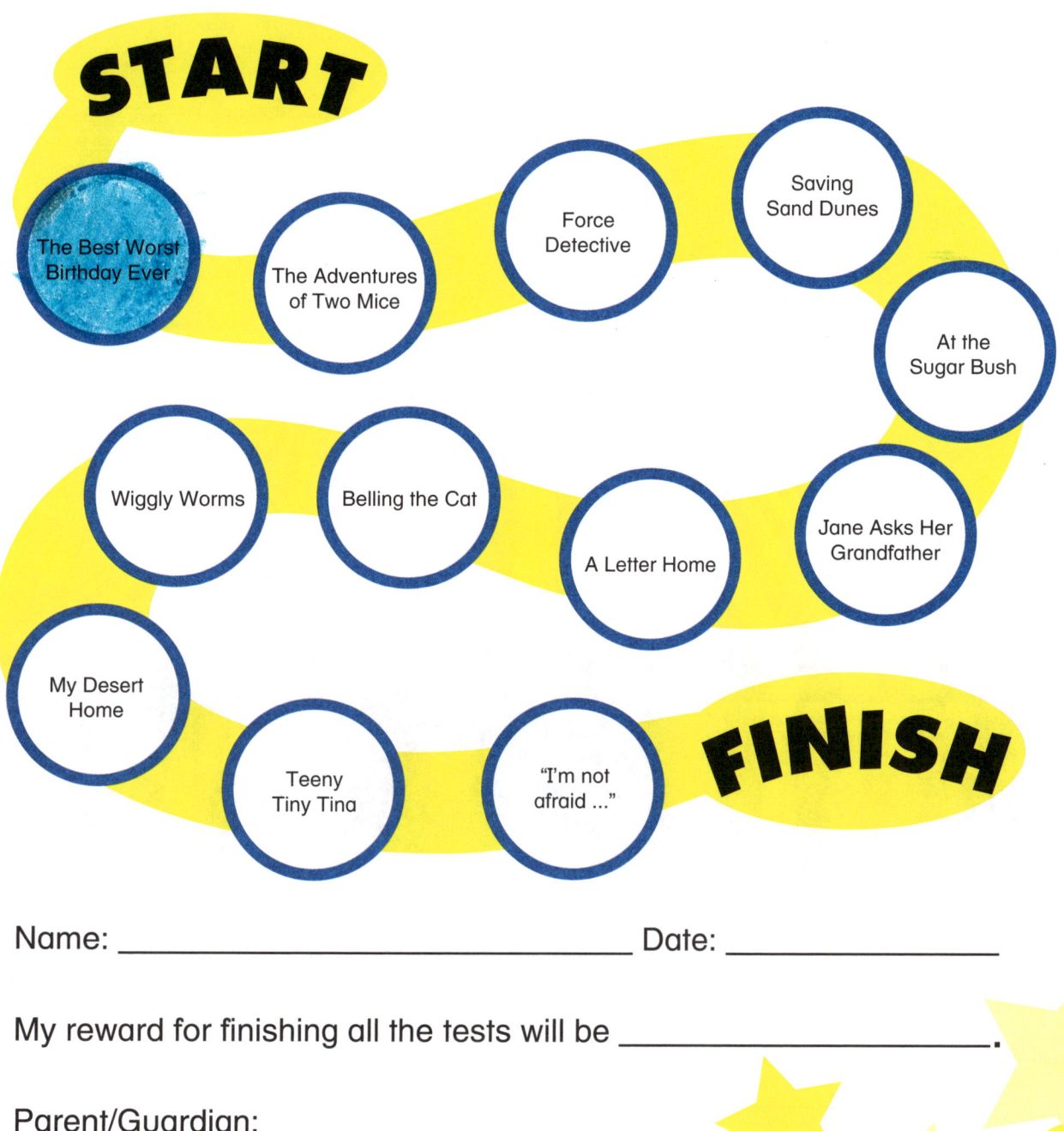

START

The Best Worst Birthday Ever

The Adventures of Two Mice

Force Detective

Saving Sand Dunes

At the Sugar Bush

Jane Asks Her Grandfather

A Letter Home

Belling the Cat

Wiggly Worms

My Desert Home

Teeny Tiny Tina

"I'm not afraid ..."

FINISH

Name: _____ Date: _____

My reward for finishing all the tests will be _____.

Parent/Guardian: _____

The Best Worst Birthday Ever

Written by Diane Robitaille

My birthday party was a disaster! Everything went wrong, but we still had a blast!

First of all, the circus tent was too big for the living room.

My father shouted at the clowns, "Turn the tent this way!"

I called out, "No, no, it will fit if you turn it that way!"

The circus ringmaster just turned his back on us and refused to help.

Finally, we got the tent inside!

"Elaine," Dad said to me, "that was way too hard. Luckily, the rest of the day should go perfectly!"

Then, the circus ponies were grumpy. They wouldn't do any tricks until we gave them some cake. Then, they complained because it wasn't chocolate!

My friend Becky kept on asking, "Where's the cake? Where's the cake?"

Finally, Dad said, "Inside the ponies!"

Josh kept telling me, "Open my present first!"

So I did, and a thousand marbles spilled across the floor.

Next, I opened Jianyi's gift. His present was shaped like a giant alligator. Guess what he got me? A giant stuffed alligator.

The alligator was bigger than our couch. I heard Dad wondering where we were going to keep it. Our apartment is pretty small.

I got lots of other great gifts too:
- one unicycle
- two goldfish with fish tank
- three stuffed dragons
- four copies of my favourite book
- a drum set

Next, the giraffe and elephant got away.

I didn't mind because they had been eating all the snacks.

I did mind when I saw that they took the loot bags with them!

"Hey, there go the loot bags!" Becky cried.

None of my friends really cared. They were busy trying on the clowns' costumes!

Nelson EQAO 3 Reading and Writing Workbook

Finally, all my friends went home.

"It's going to take *forever* to clean up this mess!"
Dad sighed.

"At least all my friends said they had a ton of fun,"
I replied.

I looked around the room and grinned.

I told Dad, "For my birthday next year, I've already invited
the zoo. I hope you like snakes!"

Reading

1. What form of writing is "The Best Worst Birthday Ever"?
   - ○ a play
   - ○ a comic strip
   - ☑ a short story
   - ○ a newspaper article

2. What problem does Elaine have with the circus tent?
   - ○ It's too small.
   - ☑ It's too big.
   - ○ It's missing.
   - ○ It's the wrong colour.

3. As it is used in paragraph 1 on page 6, the word *blast* means
   - ☑ everyone had fun.
   - ○ everyone went home hungry.
   - ○ no one arrived on time.
   - ○ they made a big mess.

4. What did the circus ponies complain about?
   - ○ the huge mess
   - ○ the big circus tent
   - ☑ the cake wasn't chocolate
   - ○ the loud children

Writing

5. Which word means the same as *grumpy*?

- ○ happy
- ☑ cranky
- ○ glad
- ○ pleased

6. Choose the word that is a compound word.

- ○ complained ?✓
- ○ living
- ○ circus
- ○ everything

7. Choose the sentence that needs a question mark.

- ☑ Where's the cake
- ○ Open my present first
- ○ Turn it this way
- ○ No, this way

8. Choose the sentence that is written correctly.
Hint: If you are not sure, check the story for this line.

- ○ Finally, all my friends went home?
- ○ Finally all my friends, went home!
- ☑ Finally, all my friends went home!
- ○ Finally all my friends went home!

Answers are on page 112.

Reading

1. Why was the birthday party a disaster?

Because first the tent would'nt fit.
Second, the ponies were grumpy.
Third, the elephant and giraffe got away.
Forth, they took the loot bags
last, it was a mess.

2. Why was it also a blast?

Because, she got gifts, tried on the
clowns costumes, and she had said,
"At least my friends had a bt tofe fun. Thereas
also was alot to do.

Writing

3. Would you like to have a party like Elaine's? Support your answer using details from the text and your own ideas.

No, because First if the tent was too big, I would feel claustrophobic. Second, clowns creep me out with thier face. Third, if a bunch of animals are running around eating our snacks, I would be very annoyed. That's why I would not like a party like Elaine's.

Answers are on page 112.

4. Describe the main character.

She always looks for the good thing in everything and everyone. She never bothered by wierd things and likes to have "Intresting" parties.

5. What do you think Elaine's next birthday party will be like? Support your answer using details from the text and your own ideas.

I think that it will be very chaotic.
1: Because the caged animals might escape
2: They might eat all the food
3: The animals might not fit
4: They are not potty-trained
5: They might eat someone
6: They live in a small apartment.
7: It will probaly be very expensive
8: Many people maye not come because they might be afraid of diffrent animals

9: They would probly need to invite a bunch of animal keepers to keep them safe. They would need money for that too

10: It would be very messy

11: What if one of her friends are claustrophobic?

Answers are on page 112.

The Adventures of Two Mice

Adapted from an Aesop's fable by Norma Kennedy

Illustrated by Barbara Spurll

Characters

MORT: a city mouse

HOWIE: a country mouse

Setting

First Act: a big city

Second Act: a small town

First Act

MORT: Welcome to my city, Howie!

HOWIE: There are so many people! Look out! We almost got stepped on!

MORT: Yes, there are lots of people here in the city.

HOWIE: What are all those huge buildings?

MORT: The really tall ones are apartment buildings and office buildings. There are also stores, banks, and restaurants on this block. There are lots of other buildings, too. Isn't it great?

HOWIE: I don't know. Where are the trees and grass?

MORT: Oh, cities have parks, too. Look, there's a park across the street. Come on! Let's go over there to get some food.

HOWIE: But there's too much traffic—buses, streetcars, cars and trucks, and bicycles!

MORT: I know! Later I'll show you the subway train. It runs underground.

HOWIE: How will we get across the street?

MORT: Just stick close to me. I'll get you there safely!

MORT: Look at all this food! Cheese, biscuits, popcorn, and cookies. All my favourites!

HOWIE: Ooohhh, my poor tummy! I've never had this kind of food. Where does it all come from?

MORT: With so many restaurants, we never run out of food.

Second Act

HOWIE: Now you'll see what it's like to live away from the city. This is a small town.

MORT: Where are the tall buildings? Why aren't there a zillion people rushing around?

HOWIE: Small towns have fewer people than big cities. This town has some of the same kinds of buildings as a city has—homes, stores, and schools. The buildings are just smaller. Look how much sky you can see!

MORT: There's not much traffic.

HOWIE: Most small towns don't have buses, streetcars, and subways. Most people walk, bike, or drive. Listen to the quiet! Smell the fresh air! We do have school buses to take the kids who live outside of town to school and home again. Here's where I live—on a farm just outside of town.

MORT: Eek! What are those monsters?

HOWIE: Don't worry about them. Cows won't bother you. But they do share some yummy food! Let's go to the barn. Try some of this tasty grain.

MORT: Oh! This doesn't taste like food! Where is all the cheese? Don't you have any cookies?

MORT: Thanks for showing me life in the country, Howie. Now I'm going back to the city, where there's good food and lots going on all the time. And no cows!

HOWIE: Thanks for showing me life in the city, Mort. I'm staying here in the country, where there's good food, lots of space, and fresh air. But let's visit again sometime soon!

Reading

1. Where does the first scene take place?

○ in a small town

○ at a farm

Ⓧ in a big city

○ at a beach

2. How do you know this reading selection is a play?

○ Because the characters and setting are identified at the beginning.

○ Because the characters take turns speaking.

○ Because the text is divided into two different acts.

Ⓧ All of the above.

3. During his visit to the small town, what scares Mort?

○ the tractors

Ⓧ the cows

○ the horses

○ the people

4. Howie usually eats

Ⓧ grain.

○ cookies.

○ cheese.

○ popcorn.

Writing

5. Which word means the same as *yummy*?

- ○ yucky
- ○ foul
- ◉ tasty
- ○ stale

6. Choose the word that uses a prefix.

- ○ truck
- ◉ bicycle
- ○ car
- ○ streetcar

7. Choose the sentence that needs a question mark.

- ○ Howie is a rural mouse
- ☑ Isn't it great
- ○ There are parks, too
- ○ It runs underground

8. Choose the sentence that is written correctly.

- ◉ How will we get across the street?
- ○ How will we get across, the street?
- ○ How will we get across, the street! ✗
- ○ How, will we get across the street?

Answers are on page 112.

Reading

1. Why do you think the author chose the title
"The Adventures of Two Mice"?

_____

_____

_____

_____

_____

2. What did Howie not like about his visit to the big city?
Support your answer with details from the text.

_____

_____

_____

_____

_____

3. Describe Mort, using details from the story and
the illustrations.

_____

_____

_____

_____

_____

_____

_____

_____

_____

_____

Answers are on page 112.

Writing

4. Where would you rather live, in a big city or a small town? Support your answer. Use the space below to plan your answer. Create an idea web or other graphic organizer.

_____

_____

_____

_____

_____

_____

_____

_____

5. Your family has spent a week at a special place. Write a journal entry describing the special place and what you did there.

_____

_____

_____

_____

_____

_____

_____

_____

_____

_____

_____

_____

_____

Answers are on page 113.

Force Detective

Nonfiction written by Marie Shah
Illustrated by Matt Roussel

"Dad! I'm home!" I shouted.

"Hi, Gumdrop. How was school?" Dad asked as he gave me a hug.

My dad calls me Gumdrop. Everyone else calls me Flora. That's my real name.

"OK, I guess. Ms. Pastin gave us homework. I have to find some examples of forces," I said.

Dad asked, "Have you thought of any?"

"Just that Ms. Pastin is *forcing* us to do this homework," I joked.

"Hey, that's pretty funny, Gumdrop. I'll leave you alone. That will *force* you to think harder," he joked back.

I rolled my eyes and then said, "I'm starving! I need a snack first."

I opened a cupboard and got down crackers and a plate. I went to the refrigerator and took out the milk and some peanut butter. Then I got a knife from the drawer and spread peanut butter on my crackers. My dog, Pudgy, stared at every move I made.

As I carried my plate to the table, Pudgy leaped up and bumped my arm. Splat! A cracker landed on the floor. Pudgy gobbled it up.

After I poured my milk, there wasn't much left.

I wrote a note for Dad to get more. Then I stuck it on the refrigerator where he would see it.

Finally, I could enjoy my tasty snack.

Now I really had to buckle down and do my homework.
I looked around like a detective searching for clues.
Where would I find examples of forces?

Suddenly, it was like a light bulb went on over my head!
In a flash, my homework was done.

When I showed it to my dad, he said he'd have to start
calling me Gumshoe instead of Gumdrop. He had
to explain to me that a gumshoe is a detective.

I think I like my new name!

CHART BY FLORA

FORCES

Muscles to Pull	Muscles to Push	Magnets to Pull	Gravity to Pull
- I pulled the cupboard door to open it. - I pulled the drawer to open it. - I pulled the fridge door to open it.	- I pushed the drawer to close it. - I pushed the fridge door to close it. - Pudgy pushed my arm.	- The fridge door uses magnets to stay closed. - I used a magnet to stick a note on the fridge.	- My cracker fell to the floor! - When Pudgy jumped up, gravity pulled him back down again.

Reading

1. In the sentence "In a flash, my homework was done."
   the phrase *in a flash* means

   ○ loudly.

   ○ sparkly.

   ⊘ quickly.

   ○ lightly.

2. Gravity is the force that

   ⊘ uses Flora's muscles to push.

   ○ pulls Pudgy back down to the floor.

   ○ makes the lights go on.

   ○ sticks a note on the refrigerator.

3. Flora leaves a note on the refrigerator

   ⊘ to remind her father to buy milk.

   ○ to remind her father to buy crackers.

   ○ to ask her father for peanut butter.

   ○ to ask her father to help her with her homework.

4. A gumshoe is

   ○ a type of candy.

   ⊘ another name for a private detective.

   ○ a type of boot.

   ○ a type of gum.

Writing

5. Which word means the same as *starving*?

- ○ full
- ○ satisfied
- ○ healthy
- ✓ hungry

6. Choose the word below that completes this sentence:
After I poured my milk, ~~There~~ wasn't much left.

- ✓ there
- ○ they're
- ○ their
- ○ they

7. Which sentence is written correctly?

- ○ Dad! I'm home! I shouted.
- ○ "Dad! I'm home! I shouted."
- ○ "Dad!" I'm home! I shouted.
- ✓ "Dad! I'm home!" I shouted.

8. Choose the word that is an adverb.

- ○ cracker
- ○ homework
- ○ forcing
- ✓ suddenly

Answers are on page 113.

Reading

1. Explain how Pudgy helps Flora with her examples of forces.

_____

_____

_____

_____

_____

2. Why does Flora roll her eyes at her father?

_____

_____

_____

_____

_____

Writing

3. What are some healthy snacks you can eat after school?

_____

_____

4. Why is it important to eat healthy snacks?

_____

_____

5. Write instructions describing how to make a sandwich.

_____

_____

_____

_____

_____

_____

_____

_____

Answers are on page 113.

Saving Sand Dunes

What Is a Sand Dune?

A sand dune is a hill made of sand. You often find sand dunes on beaches beside a lake or an ocean. Grasses and other plants often grow on sand dunes.

Sand dunes can be small like this one, or very large.

What Makes Sand Dunes?

The pulling force of wind and waves makes sand dunes.

First, moving water erodes, or breaks down, rock into tiny pieces of sand. Then, waves push sand up onto a beach. Slowly, a dune starts to form.

Wind can push more sand onto the dune, making it grow bigger.

Over time, grasses and other plants may start to grow on a dune.

Why Are Dunes Important?

Sand dunes are home to many plants and animals. Grasses and shrubs can grow on dunes. Some birds build nests in the grass on dunes. Some red foxes make their homes among dunes.

Sand dunes are essential for many different animals.

Red foxes hunt for food on sand dunes.

What Threatens Sand Dunes?

A big storm can destroy a sand dune. Each time a wave hits a dune, it pulls some of the sand back into the water. Lots of big waves can wash away the whole sand dune.

People are also dangerous to sand dunes. When people walk on the dunes, they may kill the plants growing there. The roots of the plants help to keep the sand in place.

Also, when people walk on dunes, they might scare away animals that live there.

Walking on dunes can kill the plants that grow there. In many areas with sand dunes, people construct boardwalks over the dunes.

How Can People Protect Sand Dunes?

Here are some ways to protect sand dunes:

1. Put large branches on the dunes. These branches can help stop the wind from pushing sand off the dunes.

2. Build wooden paths through the dunes. These walkways can stop people from walking on and killing the plants.

3. Put up fences or signs to keep people away from the dunes.

Reading

1. When a storm hits a sand dune,

○ some of the sand gets pulled into the water.

○ the wet sand makes the dune stronger.

○ the foxes climb up to the top of the dunes.

○ the dry sand on top gets hard.

2. In paragraph 3 on page 36, the term *erodes* refers to

○ turning the sand into water.

○ making the rocks stronger.

○ breaking the rocks down into sand.

○ making the rocks shiny.

3. How do branches help sand dunes?

○ Branches erode and become sand.

○ Little trees grow from the branches.

○ Branches stop the wind from pushing sand off the dunes.

○ Birds build nests in the branches.

4. Why are plants important for sand dunes?

○ Plant roots help keep the sand in place.

○ Plant leaves help protect sand from the wind.

○ Plant stems soak up rainwater.

○ Plant flowers attract bees.

Writing

5. Which word means the same as *destroy*?

○ aid

○ damage

○ assist

○ help

6. Choose the sentence that is written correctly.

○ Slowly a dune, starts to form.

○ Slowly a dune starts, to form.

○ Slowly a dune starts to form

○ Slowly, a dune starts to form.

7. Choose the right word(s) to complete this sentence:
Some red foxes make _____ homes among dunes.

○ their

○ there

○ they're

○ they are

8. Choose the plural word.

○ pushing

○ walkways

○ person

○ fence

Answers are on page 113.

Reading

1. Describe how sand dunes are formed.

_____

_____

_____

_____

_____

_____

_____

2. Is "Saving Sand Dunes" fiction or nonfiction?
   Support your answer using information from
   the text and your own ideas.

nonfiction Beacuase it is a Real
Story and if teachis us about things for
when we grow alder.

_____

_____

_____

Writing

3. Describe a time when you helped the environment.

_____

_____

_____

_____

_____

_____

4. Write a journal entry about spending the day at the beach. Think about the following questions: What did you do at the beach? How long were you there? Who did you go with?

_____

_____

_____

_____

_____

Answers are on pages 113 to 114.

At the Sugar Bush

Written by Benoît Leclerc

Gran wakes me up while it is still dark outside. "Time to get up, Abbey!" Gran hollers from downstairs.

We eat breakfast and watch an orange glow spread across the sky. The sun is rising and there are no clouds. It's going to be a perfect day to make maple syrup.

For March break, I'll spend all week at Gran's. I'm helping her with sugaring off—that's what we call collecting sap from the maple trees and making maple syrup.

We put on warm clothes and walk out to the sugar bush. This is where the sugar maple trees grow. Almost every tree has a bucket attached to it to collect sap.

In the snow, two long trails of footprints stretch out behind us. I get to pull the sled. The sled has also made two long lines in the snow.

At the sugar bush, we find the trees that have a pail hanging from a metal tube. A few days ago, Gran went out and gently hammered the tubes into the trees so the sap would drip out. I help Gran empty the pails of sap into a big barrel.

Gran starts a fire under the big sap pan.

Then, she pours in the sap. Soon, sweet-smelling steam comes curling out of the pan as the sap boils.

Under the pan, the coals of the fire glow red-hot. We pour more sap into the pan.

After the sap boils for a long time, it becomes a thick, brown syrup. I help Gran hold the cloth as she pours the syrup through it. The cloth catches any dirt and ash in the syrup.

The sun is setting as we head for home. Long shadows of trees fall across the snow. Gran pulls the sled with the heavy pails of maple syrup. We're both happy when we see the lights of the farmhouse twinkling in the distance.

Tomorrow, Gran will make my favourite breakfast. We'll have pancakes with delicious maple syrup on top!

Reading

1. When is sap collected to make maple syrup?

○ early summer

✓ late winter *late winter*

○ late spring

○ early fall ✳

2. Gran boils the sap in

○ a large pot with a heavy lid.

○ metal tubes and buckets.

✓ a sap pan.

○ a large barrel.

3. Why does Gran need a cloth when she makes maple syrup?

✓ The cloth is used to filter dirt and ash out of the syrup.

○ The cloth is used to clean all of the trees.

○ The cloth is used to collect the sap from the trees.

○ The cloth is used to cover all of the pails.

4. What does the word *it* refer to in this sentence: After the sap boils for a long time, it becomes a thick, brown syrup?

○ the ash

○ the sap

○ the dirt

○ the syrup

Writing

5. Abbey says that she helps Gran empty the pails. Which word means the same as *empty*?

- ○ fill
- ☑ pour out
- ○ pack
- ○ lift

6. Choose the word below to complete the following sentence: Gran wakes me up _While_ it is still dark outside.

- ☑ while
- ○ after
- ○ during
- ○ if

7. Choose the sentence that uses the correct contraction for the words *it is*.

- ○ Its Gran's favourite time of year.
- ○ Its' very cold outside.
- ☑ It's going to be a perfect day to make maple syrup.
- ○ I'ts getting wet on the ground.

8. Choose the word that is a proper noun.

- ○ farmhouse
- ☑ Abbey
- ○ distance
- ○ sap

Answers are on page 114.

Reading

1. Describe the process of making maple syrup.

_____

_____

_____

_____

_____

2. How do you know that Abbey likes being
in the sugar bush?

_____

_____

_____

_____

Writing

3. What is your favourite season? Support your answer.

_____

_____

_____

_____

_____

4. Your teacher wants to take the class on a field trip. Write a letter explaining to your teacher where you think the class should go.

_____

_____

_____

_____

_____

_____

_____

_____

Answers are on page 114.

Jane Asks Her Grandfather

Hi! I'm Jane Wolf. I belong to the Squamish First Nation. My people were in Canada long before the pioneers arrived. For a school project, I asked my grandfather about how our people lived long ago.

The first thing I wanted to know was what foods our people ate. Instead of answering, Grandfather asked *me* a question: "What did you eat for dinner last night?"

"Umm … wild rice, fish, and squash," I said. "And pumpkin pie. My favourite!"

Grandfather laughed. "All the foods you ate were the same foods that Indigenous Peoples enjoyed hundreds of years ago," he said. "Popcorn and maple syrup came from Indigenous Peoples, too."

Many Indigenous Peoples called corn, beans, and squash the "Three Sisters." These three plants grow together well. The beans climb up the cornstalks, and the broad squash leaves stop weeds from growing.

The next thing Grandfather talked about was tools.

"Did you know that Indigenous Peoples didn't use metal to make their tools?" he said.

I was surprised. Today, we use metal for almost every tool.

Grandfather told me that Indigenous Peoples made sharp knives, drills, and arrows from stone, wood, and bone.

These stone tools were made hundreds of years ago.

"My mom drives us to the mall. But long ago, our people used canoes to travel, right?" I asked.

Grandfather nodded. "We made canoes out of birchbark. Our canoes were strong and could carry big loads."

Grandfather smiled and added, "Now, we have canoes made out of fibreglass or aluminum. Remember how we went fishing last summer in an aluminum canoe? It was light enough to carry over the trails."

Grandfather looked thoughtful for a moment and then said, "In winter, we used snowshoes to walk in deep snow. We still do. I can take you snowshoeing next winter."

Snowshoes stop you from sinking in deep snow.

Indigenous Peoples made canoes from birchbark because the bark was light and easy to carry. Today, canoes are still made from birchbark. But other materials are also used, such as aluminum and fibreglass.

Next, I wanted to know about the Elders.

Grandfather told me that Elders played a very important role in the community, just like they do today.

"Elders were chosen because they were good at listening to people," he explained. "The people went to Elders when they needed advice. The Elders were teachers, too."

"I'm learning from you right now," I said.

"I'm glad to hear that, Jane," said Grandfather, smiling.

Elders pass along teachings to young people. This is a photo of my grandfather with my cousins. He's always happy to teach us!

Reading

1. Why did Jane Wolf interview her grandfather?
   - ○ She wanted to learn about canoes.
   - ○ She was doing a project for school.
   - ○ She liked hearing stories about animals.
   - ○ She was writing a short play.

2. Snowshoes help people to
   - ○ walk over deep snow.
   - ○ walk quietly when hunting animals.
   - ○ float on the water.
   - ○ scare away large bears.

3. Why was Jane surprised about what her Grandfather said about tools?
   - ○ She thought metal was needed to make tools.
   - ○ She thought drills needed electricity.
   - ○ She thought people didn't have tools long ago.
   - ○ She thought people bought tools at stores.

4. Why did Grandfather smile at the end of the story?
   - ○ He was happy he and Jane liked the same types of food.
   - ○ He enjoyed telling stories about canoeing.
   - ○ He was pleased that Jane was learning new things.
   - ○ He thought Jane was very brave.

Writing

5. Which word means the opposite of *laughed*?

○ cackled

○ snickered

○ chuckled

○ cried

6. Choose the sentence that is written correctly.

○ The first thing I wanted too know was what foods our people ate.

○ The first thing I wanted two know was what foods our people ate.

○ The first thing I wanted 2 know was what foods our people ate.

○ The first thing I wanted to know was what foods our people ate.

7. Choose the right word to complete the following sentence: Elders were chosen _____ they were good at listening to people.

○ so

○ because

○ until

○ again

8. Choose the word that is an adjective.

○ popcorn

○ pioneers

○ elders

○ happy

Answers are on page 114.

Reading

1. What does Jane learn from her grandfather?

_____

_____

_____

_____

_____

_____

2. Why did Indigenous Peoples use birchbark to build canoes?

_____

_____

_____

_____

_____

Writing

3. Why are Elders important in a community?

_____

_____

_____

_____

_____

4. This selection uses a then-and-now comparison structure. Ask an older family member about their childhood. Write a then–now paragraph comparing their childhood to yours.

_____

_____

_____

_____

_____

_____

Answers are on page 114.

A Letter Home

Historical Fiction by Kevin Carruthers

November 18, 1799

Dear Cousin Andrew,

At last I have time to write! Life is very busy right now. We have to harvest the crops and get ready for our first winter in Upper Canada.

Even though we have been here a while, I'm still surprised at all the trees on our land. It's so different from Scotland's bare, stony hills.

I know your parents don't want to leave Scotland, but I think you would like it here.

We had a logging bee. All of our neighbours came to help us build a snug log cabin.

Mary, John, and I collected stones so Father could build a big fireplace. We planted potatoes, corn, and a little wheat.

My daily chores are keeping the woodbox filled and getting buckets of water from the stream. I also have to feed the pigs.

Yesterday, Mary and I gathered nuts from the forest.
We'll use them to fatten up the pigs we're raising.

Our neighbour will show us how to salt the meat
for the winter and smoke the hams in our chimney.

Today, Father and John took our wheat to the mill.
We have no wagon or horse yet, so they had to carry
the sacks on their backs. It will be good to have
fresh bread from our own flour.

Mother, Mary, and I dug potatoes all day. The work here is harder than at home.

Mary is so small. She cries because her shoulders hurt from churning butter.

Mother says we're all toughening up. Having fresh butter on our own home-baked bread is worth all the hard work! She says it's good to be working on our own land.

Are you still thinking of joining us? Mother says your skill as a weaver would be welcome. All the farms around us have sheep, so there is a lot of wool for weaving. I will say goodbye for now. Please let the rest of the family know how much we miss them.

Your cousin,

Jamie McLeod

Reading

1. Jamie, the writer of the letter, lived in Upper Canada. His cousin, Andrew, lived

   ○ in Upper Canada.

   ○ in Lower Canada.

   ○ in the United States.

   ○ in Scotland.

2. Jamie's daily chores were

   ○ feeding the neighbour's sheep and pigs.

   ○ churning the butter and collecting wood.

   ○ keeping the woodbox filled and getting buckets of water.

   ○ hunting rats.

3. In the first paragraph on page 61, the word *bee* means

   ○ an insect that makes honey.

   ○ a group of people working together on a project.

   ○ the second letter of the alphabet.

   ○ a log shaped like a bee.

4. What did Jamie, his mother, and Mary do all day?

   ○ They relaxed in the sun.

   ○ They dug potatoes.

   ○ They picked strawberries.

   ○ They played with the sheep.

Writing

5. Which word means the same as *harvest*?

○ gather

○ eat

○ ignore

○ burn

6. Choose the correct word to complete the sentence.
She cries _____ her shoulders hurt from
churning butter.

○ so

○ again

○ until

○ because

7. Which sentence is written correctly?

○ We had a logging bee, so we were able to build a snug log cabin.

○ We had a logging bee, so we was able to build a snug log cabin.

○ We had a logging bee, so we where able to build a snug log cabin.

○ We had a logging bee, so we's able to build a snug log cabin.

8. Choose the word that uses a suffix.

○ weaver

○ chimney

○ neighbour

○ goodbye

Answers are on page 114.

Reading

1. Explain why Jamie's mother said, "We're all toughening up." Support your answer with details from the text and your own ideas.

_____

_____

_____

_____

_____

_____

2. Why do you think Jamie wrote a letter to his cousin?

_____

_____

_____

_____

_____

_____

Writing

3. Using ideas from "A Letter Home," explain what you would like about being an early settler.

_____

_____

_____

_____

_____

_____

4. Now, explain what you would dislike about being an early settler.

_____

_____

_____

_____

_____

_____

Answers are on page 115.

Belling the Cat

Adapted from an Aesop's fable by Rosanne Czuba

Mice and cats have been enemies forever.

After all, it is a cat's job to chase, catch, and then eat mice.

In one farmhouse, the mice had been living for several years without any worries. The farmer kept the cat in the barn. There, the cat hunted the rats.

Then one day, all the rats were gone.

The farmer decided to bring the cat into the house.

The mice met to talk about the problem.

A young mouse complained, "Every time I poke my head out of a hole, the cat is there."

"Something has to be done," another mouse said.

"The cat chases us one at a time," pointed out an old mouse. "Maybe if we all chased the cat, we would scare it."

The grandmother mouse shook her head and said, "Well, we might frighten it at first. But after a while it would be back. And it would be angry!"

She was a very wise mouse.

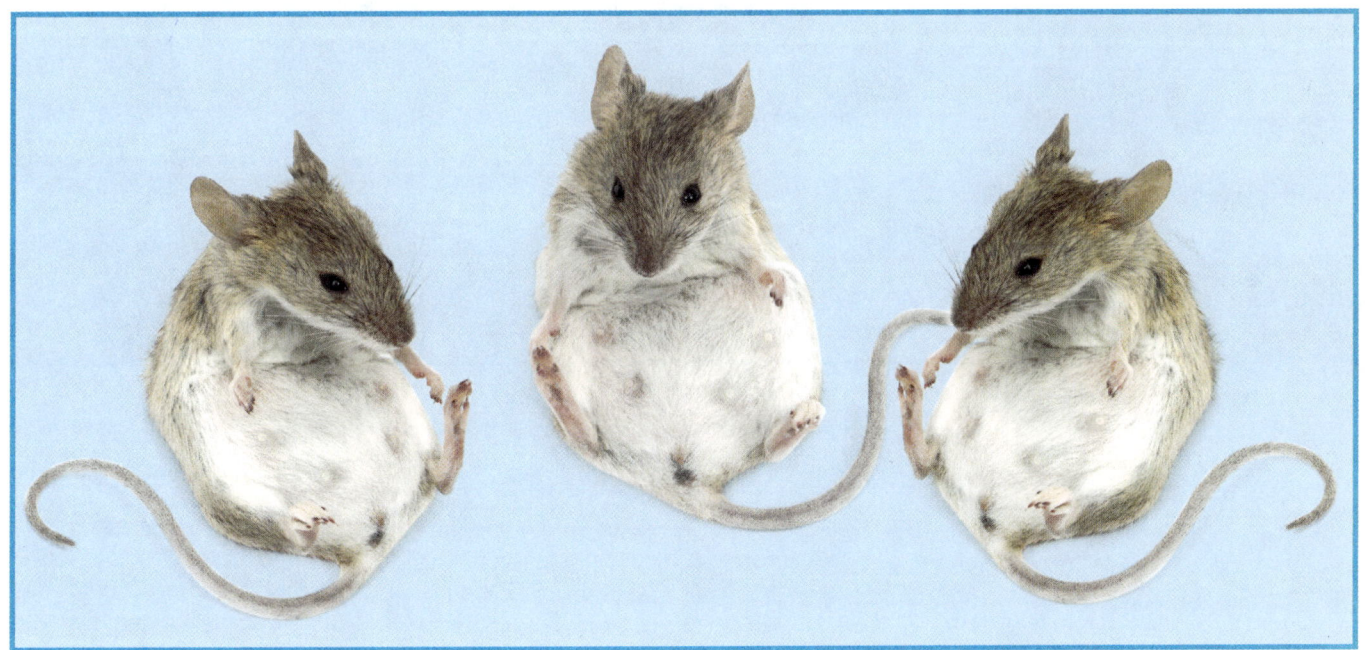

"You're right!" agreed the youngest mouse. "But if we could hear the cat coming, we could run away quickly."

"We could put a bell on the cat's collar. That way, we will always know where it is," suggested the first mouse.

Most of the mice thought belling the cat was a great idea. They knew they'd never have to worry about the cat again if it was wearing a bell. All the younger mice hopped around happily.

But the grandmother mouse shook her head. "Which mouse is brave enough to put a bell on the cat?" she asked.

Moral: Talk is easier than action.

Reading

1. This story is a fable. A fable is

○ a true story with a moral.

○ usually nonfiction.

○ a tale with a moral.

○ always about cats.

2. The mice had been living in the farmhouse without any worries because

○ the cat slept in the farmer's room.

○ the cat lived in the barn.

○ the cat was afraid of the grandmother mouse.

○ cats and mice have been friends forever.

3. Why does the grandmother mouse shake her head in the last paragraph?

○ She knows no mouse will be brave enough to bell the cat.

○ She thinks belling the cat is a clever idea.

○ She knows they would never have to worry about the cat again.

○ She thinks it will make the cat angrier.

4. Why do the mice think belling the cat is a good idea?

○ The noise of the bell will scare the cat.

○ The cat will thank them for the gift.

○ The bell will signal when the cat is nearby.

○ The cat will want to make a truce.

Writing

5. Which word means the same as *complained*?

○ grumbled

○ praised

○ yelled

○ commented

6. Choose the sentence that uses the correct contraction for the words *you are*.

○ "Your'e right!" agreed the youngest mouse.

○ "Youre right!" agreed the youngest mouse.

○ "You're right!" agreed the youngest mouse.

○ "Y'ure right!" agreed the youngest mouse.

7. Choose the sentence that is punctuated correctly.

○ After all, it is a cat's job to chase, catch, and then eat mice.

○ After all it is a cat's job to chase, catch, and then eat mice.

○ After all, it is a cat's job to chase catch, and then eat mice.

○ After all, it is a cat's job to chase catch and then eat mice.

8. Choose the word that is a verb.

○ hunter

○ frighten

○ youngest

○ farmhouse

Answers are on page 115.

Reading

1. The moral of this story is "Talk is easier than action."
Explain what this moral means. Use details from
the text and your own ideas.

_____

_____

_____

_____

_____

2. Why did the farmer move the cat into the house?

_____

_____

_____

_____

Writing

3. Rewrite this story from the cat's point of view.

I have been eniemies w

Answers are on page 115.

Wiggly Worms

Written by Aaron Meleski

Hi, I'm an earthworm. You've probably seen me and my friends in your garden, on your grass, or even on the sidewalk after a rainy day. Please don't step on me. Earthworms do great things for the soil.

It was really sunny out today, so I stayed deep under the vegetable garden. I don't like the sun. It dries up my skin. I said "hi" to a few friends and then got to work slithering through the soil, mixing the soil layers together.

Castings help soak up water that the soil needs for plants to grow.

I was really hungry today.

I munched on soil, dead leaves, and insects. I gobbled all that tasty food up with my mouth.

From there, the food slid into my crop, and then down to my gizzard. I don't have any teeth to chew my food with. I have muscles that help me digest my food.

After I finished eating, little pellets called *castings* came out of my tail. Castings make great food for plants because they are packed with nutrients and minerals.

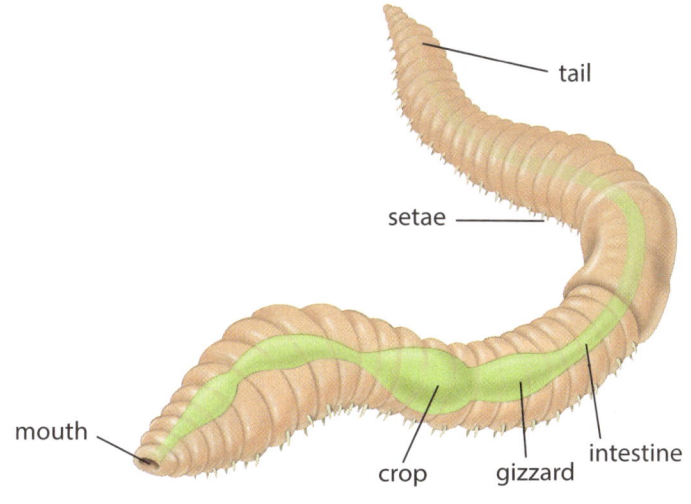

tail

setae

mouth

crop

gizzard

intestine

An earthworm's digestive system has many parts.

Worms like to eat dead leaves with pointed ends.

I had a fright near the end of the day. A little girl dug me up with her shovel! I wriggled off the shovel and into the dirt before she saw me. I took some bits of dead grass and leaves as I went.

It's not safe for worms to be out of the soil for too long.

We need the moist soil to help us breathe. The only time we come out is at night and after it rains.

One time, a worm friend of mine had a tiny piece of his tail cut off. He was OK. He grew another tail. That was possible because our bodies are made out of segments. We worms can keep living if we have enough segments left.

Phew! It's a lot of work twisting, tunnelling, and eating through the soil. Did you know that I eat my weight in soil every day? It's true. The hard work is worth it. I help the soil grow healthy flowers and vegetables.

There may be as many as 30 hard-working worms in a bucket of good garden soil.

Reading

1. Worms have
- ○ sharp teeth.
- ○ bodies made of segments.
- ○ three eyes.
- ○ long tongues.

2. Looking at the diagram of the worm, *setae* are
- ○ muscles.
- ○ the stomach.
- ○ shoulders.
- ○ small hairs.

3. What happens when a worm is exposed to too much sun?
- ○ It gets a sunburn.
- ○ Its skin dries up.
- ○ It can't find food.
- ○ It gets thirsty.

4. The first two sentences in the last paragraph on page 79 explain
- ○ why the worm is tired.
- ○ why the worm is angry.
- ○ why the worm is bored.
- ○ why the worm is worried.

Writing

5. Which word means the same as *fright*?

○ calm

○ happy

○ scare

○ content

6. Choose the sentence that is written correctly.

○ It's a lot of work twisting, tunnelling, and eating through the soil.

○ It's a lot of work twisting tunnelling, and eating through the soil.

○ It's a lot of work twisting tunnelling and eating through the soil.

○ Its a lot of work twisting, tunnelling and eating through the soil.

7. Choose the right word to complete the following sentence: It was really sunny out today, so I stayed deep _____ the vegetable garden.

○ over

○ beside

○ under

○ along

8. Choose the word that is a singular noun.

○ castings

○ earthworm

○ gobbled

○ shovels

Answers are on page 115.

Reading

1. Describe what worms eat. Then, explain how they digest food.

_____

_____

_____

_____

_____

2. How are worms helpful to plants? Support your answer with information from the text and your own knowledge.

_____

_____

_____

_____

_____

Writing

3. Your class has decided to plant a school garden.
Write a letter to your teacher explaining what you
think should be included in the garden.

_____

_____

_____

_____

_____

_____

_____

_____

_____

_____

_____

_____

Answers are on pages 115 to 116.

My Desert Home

Hi. My name is Ada. I live in the Sahara Desert with my family.

Our people, the Tuareg, have lived in the desert for thousands of years. We know all about the desert!

Lots of people think they know what it must be like to live in a desert. Some think deserts are nothing but sand dunes. Other people think plants don't grow in the sandy soil, or animals can't live in the sand.

Wow, are they wrong!

Ada's family cooks food over a campfire in the desert.

More Than Just Sand Dunes

The Sahara Desert has more rocks than sand, but we do have a lot of sand. One sand dune in the Sahara is as large as France!

One interesting fact about our sandy soil is that it's very salty. Thousands of years ago, the Sahara was covered by seawater!

Now, there's hardly any water at all.

Sand in the Sahara Desert doesn't have a lot of organic matter. Organic matter is the dead plants and animals that make soil rich.

Desert Plants

The Sahara Desert is very hot and dry. So, many plants that grow in Canada will not grow here. But there are different types of plants that will grow in the desert.

Most plants that grow in the desert have long roots. Long roots can reach the water that is far underground. Some plants store water in their stems. Other plants only begin to grow when it rains.

Most desert plants have seeds that can live in the soil for years, just waiting for water before they grow.

The Sahara Desert has oases, places where underground rivers come close to, or above, the surface. Plants can grow well in the soil of an oasis.

If you pick up a handful of desert sand, you can see seeds from different plants.

Desert Animals

Many animals live in my beautiful desert. My family keeps camels, goats, sheep, and donkeys. Wild animals also live in the desert. These are just a few of the animals that live here: scorpions, lizards, foxes, gerbils, and cobras and other snakes.

Many desert animals, like bugs, snakes, and the sand cat, keep cool by staying underground during the day. The deeper you dig down into the sandy soil, the cooler it is!

My desert is hot, dry, sandy, and beautiful!

Desert mole rats live in burrows deep in the sandy soil. They use their front teeth to dig burrows. For food and water, mole rats eat plants and roots.

Reading

1. Desert mole rats live in burrows. What is a burrow?

- ○ an oasis
- ○ a hole
- ○ a pile of leaves
- ○ a sand dune

2. A sand cat stays underground during the day

- ○ to hide from snakes.
- ○ to collect salt.
- ○ to keep cool.
- ○ to hunt rats.

3. In paragraph 2 on page 84, the word *Tuareg* starts with a capital letter because it is

- ○ a proper noun.
- ○ at the beginning of a sentence.
- ○ important for the reader to know.
- ○ the title of a story.

4. What does this paragraph explain?
Many desert animals, like bugs, snakes, and the sand cat, keep cool by staying underground during the day. The deeper you dig down into the sandy soil, the cooler it is!

- ○ It explains how desert animals survive in the heat.
- ○ It explains what types of pets Ada has.
- ○ It explains why desert animals are dangerous.
- ○ It explains where you can find camels.

Writing

5. Which word means the same as *hardly*, as it is used in this sentence: Now there's hardly any water at all.

- ○ very
- ○ greatly
- ○ frequently
- ○ barely

6. Choose the word that best fills in the blank:
Some plants store water in _____ stems.

- ○ there
- ○ they're
- ○ their
- ○ they

7. Choose the sentence that is written correctly.

- ○ Most plants that grow in the desert have long roots.
- ○ Most plant that grow in the desert have long roots.
- ○ Most plants that grow in the desert have long root.
- ○ Most plant that grow in the desert have long root.

8. Choose the word that best fills in the blank:
_____ food and water, mole rats eat plants and roots.

- ○ Of
- ○ In
- ○ By
- ○ For

Answers are on page 116.

Reading

1. Why does Ada say, "Wow, are they wrong!" in paragraph 4 on page 84? Support your answer with details from the text and your own ideas.

_____

_____

_____

_____

_____

2. How do plants survive and grow in the Sahara Desert? Support your answer with details from the text and your own ideas.

_____

_____

_____

_____

_____

Writing

3. Describe some of the characteristics of the sand in the Sahara Desert.

_____

_____

_____

_____

_____

4. How do you know Ada is proud of where she lives?

_____

_____

_____

_____

_____

_____

Answers are on page 116.

Teeny Tiny Tina

Written by Diane Robitaille

Once upon a time, in a teeny tiny village in a teeny tiny house, there lived a teeny tiny woman named Tina.

All her friends called her Tiny Tina.

All her enemies called her Tiny Tina the Terrible.
More about how terrible she could be later …

Tina had tiny feet and tiny hands and a tiny face with a tiny nose and tiny eyes and teeny tiny teeth, each one no bigger than a grain of sand.

However, Tiny Tina's voice was a GREAT BIG VOICE. More about her GREAT BIG VOICE later …

Tina was so tiny that she could take a bath in a teacup.

She could wear a handkerchief as a coat.

Tiny Tina could use the top of an acorn as an umbrella.

All the groceries she would need for a week could fit in a soup bowl.

Tina was so tiny that she—and all her luggage—could travel on a bus or an airplane in someone's coat pocket. She would crawl into a little round jar that was capped with a lid with a few holes in it.

Tina saved a lot of money by living this way.

Of course, those are the positives.

There were a lot of negatives to being that tiny.

For example, sometimes, people forgot Tiny Tina was even there.

Sometimes, people might even have almost stepped or sat on Tiny Tina, if they hadn't seen her first.

That's why Tiny Tina had such a GREAT BIG VOICE. So that she could yell, "HEY, YOU BIG OAF! CAN'T YOU SEE ME? I'M RIGHT HERE!!!"

Tiny Tina could be a bit rude sometimes. In fact, terribly rude. But I guess if someone almost crushes you, you might be moved to rudeness!

Think about it this way. If a regular-sized person stepped or sat on Tiny Tina, it would be like a great big elephant almost stepping or sitting on me or you. I definitely wouldn't want that to happen, would you?

With her GREAT BIG VOICE, Tiny Tina could say some TERRIBLE things. For example, she never let anyone get away with littering. "HEY!" she would holler, "THIS IS MY PLANET, TOO!"

Tina *was* sometimes terribly blunt. So, she did make a few enemies.

However, Tiny Tina also made some wonderful friends. Friends who loved her GREAT BIG VOICE as much as they loved her teeny tiny self.

They loved her because Tiny Tina was funny and loved to laugh. And she had a GREAT BIG LAUGH that made everyone laugh along with her.

They loved her because Tiny Tina was smart and knew just what to do in any situation.

Finally, most especially, people loved Tiny Tina because Tiny Tina's heart *wasn't* teeny tiny. Tiny Tina was very generous. All the money she saved on food, clothes, and travel she gave away to people in need—family, friends, or strangers.

Tiny Tina's GREAT BIG VOICE helped her make a good living as a radio-show host. So, she made *stacks* of money. Stacks of money that were taller than Tiny Tina.

Tiny Tina also made lots of fantastic speeches in her GREAT BIG VOICE—about the importance of being kind to people and the environment.

Her best speech, of course, was about being proud of who you are.

As she liked to say, "I'M TINA AND I'M TINY!! I MIGHT BE TEENY TINY AND YOU MIGHT ALL BE GREAT BIG LUMPS THAT BLOCK OUT THE SUNSHINE, BUT WE ALL HAVE TO LIVE TOGETHER AND LEARN TO GET ALONG! SMALL OR TALL, LOVE 'EM ALL!"

Reading

1. How would you describe Tina?
   - ○ a tall, quiet woman with no personality
   - ○ a young woman who works as an elephant trainer
   - ○ an angry woman who is afraid of other people
   - ○ a small, clever woman with a great big voice who loves to help others

2. What does Tina use as a bathtub?
   - ○ a soup bowl
   - ○ a teacup
   - ○ an acorn top
   - ○ a small jar

3. As it is used in the exclamation, "HEY, YOU BIG OAF!" the word *oaf* means
   - ○ a loaf of bread.
   - ○ someone clumsy.
   - ○ someone silly.
   - ○ a big rock.

4. Tina likes being teeny tiny because
   - ○ people sometimes step on her.
   - ○ she saves a lot of money that way.
   - ○ people sometimes don't even see her.
   - ○ she can use a handkerchief for a coat.

Writing

5. What is the main purpose of paragraph 1 on page 92?

○ It introduces the main character and the setting.

○ It introduces the problem in the story.

○ It resolves the problem in the story.

○ It introduces the setting.

6. In the sentence "SMALL OR TALL, LOVE 'EM ALL!" the apostrophe replaces one or more letters. What word was the writer thinking of?

○ hem

○ stem

○ rem

○ them

7. Which sentence is punctuated correctly?

○ "HEY, she would holler, THIS IS MY PLANET, TOO!"

○ "HEY!" she would holler, "THIS IS MY PLANET, TOO!"

○ "HEY," she would holler, "THIS IS MY PLANET, TOO"

○ "HEY!" She would holler. THIS IS MY PLANET, TOO!"

8. Choose the word that best fills in the blank in the following sentence: They loved her _____ Tiny Tina was smart.

○ since

○ if

○ because

○ when

Answers are on page 116.

Reading

1. Describe how Tina lives.

_____

_____

_____

_____

_____

_____

2. Why does the writer describe Tiny Tina as "terribly blunt"?

_____

_____

_____

_____

_____

3. Do you think you would like Tiny Tina? Support your answer with details from the text and your own ideas.

_____

_____

_____

_____

_____

_____

_____

_____

_____

_____

_____

_____

Answers are on page 116.

Writing

4. Describe what you would like about being as teeny tiny as Tina.

_____

_____

_____

_____

_____

_____

5. What do you think one of Tina's enemies would say about her?

_____

_____

_____

_____

_____

_____

_____

6. Write a letter to Tina. Explain how you feel about her speech in the last paragraph. Check your spelling, grammar, and punctuation.

_____

_____

_____

_____

_____

_____

_____

_____

_____

_____

_____

_____

_____

Answers are on page 117.

"I'm not afraid ..."

Written by Philippe Levesque

Bryce stood well back from the ledge as his father stepped closer.

"Come on, Bryce! Don't be afraid," Dad called back.

"I'm not afraid," Bryce insisted. Bryce's father was always coaxing him to do stuff.

"An extreme fear of heights, also known as *acrophobia*, is quite common. Nothing to be ashamed of," Bryce's mother explained. "*Phobia* comes from the Greek word for *fear*. *Acro* comes from the Greek word for *summit*."

Bryce rolled his eyes. His mother was always explaining things.

"I'm *not* afraid of heights," Bryce repeated. He paused, then added, "I'm *afraid* of falling!"

He stood well back and took a photo of his parents. The view *was* fantastic, but he was as close as he needed to be.

Walking back down the mountain, Bryce leaped back from the sudden appearance of a spider on the trail.

"Don't be afraid, Bryce. It's just a spider," his dad whispered, stepping closer. "She's more afraid of you than you are of her."

"I'm not afraid of spiders," Bryce said backing away.

"Arachnophobia," his mother explained, "is an extreme fear of spiders. *Arachne*, of course, is the Greek word for *spider*."

"I'm *not* afraid of spiders," Bryce repeated. "I'm *afraid* of spider bites."

It *was* a gorgeous spider. Bryce had never seen one like it before.

Bryce used his zoom lens to take a photo.

Walking on, they reached a point in the path that brought them back in sight of the river.

On the other side of the river was a bear catching a salmon!

"AHH!" Bryce cried in surprise.

"It's OK, Bryce, we're far enough away! That bear won't bother us," his dad said soothingly.

"I'm not afraid of the bear! I was just surprised," Bryce protested.

"Agrizoophobia," his mother explained, "is an extreme fear of wild animals. *Agri* is the Greek word for *field*. *Zoo* is the Greek word for *animals*."

"I'm *not* afraid of wild animals," Bryce repeated. "I'm *afraid* of being chased by wild animals."

Bryce snapped a photo quickly and then started walking rapidly down the path. The path turned again, taking them farther away from the hungry bear.

That night, Bryce and his parents were settling down in their sleeping bags. The door of the trailer was still open. It was incredibly dark in the forest. The trees cut out any light from the stars and the moon. The animals outside were making a LOT of noise.

Bryce was fiddling with his camera.

"Don't be afraid of the dark, Bryce. We're right here," his dad said reassuringly.

"I'm not afraid of the dark," Bryce protested. He was examining all the photos he had taken that day.

Outside, a wolf howled.

His parents exchanged glances. *"We're* sort of afraid! It's OK if you are."

Bryce looked up from his photos and grinned. "Achluophobia," he explained, "is an extreme fear of darkness. *Achluo* is the Greek word for *darken*."

Bryce got up and closed the door. "I'm not afraid of the dark," he repeated. "YOU are! Good night. Sleep tight." Bryce smiled and went to sleep.

Reading

1. How does Bryce feel about spiders?
   - ○ Spiders scare him.
   - ○ Spiders fascinate him.
   - ○ Spiders annoy him.
   - ○ Spiders make him angry.

2. What is agrizoophobia?
   - ○ a fear of heights
   - ○ a fear of wild animals
   - ○ a fear of spiders
   - ○ a fear of darkness

3. Bryce and his parents are
   - ○ hiking and camping in the mountains.
   - ○ visiting family.
   - ○ sightseeing in a big city.
   - ○ driving across Canada to move to another town.

4. Choose the sentence that is true.
   - ○ Bryce likes to take photos.
   - ○ Bryce doesn't like bears.
   - ○ Bryce's parents think he isn't afraid of anything.
   - ○ Bryce's parents are afraid of heights.

Writing

5. Which word means the same as *reassuringly*?

○ angrily

○ soothingly

○ happily

○ joyfully

6. In which sentence is the exclamation mark used correctly?

○ "I'm *afraid* of falling!"

○ "I'm *afraid* of falling"!

○ "I'm *afraid*! Of falling!"

○ "I'm *afraid* of falling!"!

7. Which sentences are punctuated correctly?

○ "I'm not afraid of the bear! I was just surprised" Bryce protested.

○ "I'm not afraid of the bear! I was just surprised," Bryce protested.

○ "I'm not afraid of the bear I was just surprised," Bryce protested.

○ "I'm not afraid of the bear! I was just surprised, Bryce protested."

8. Reread the following paragraph.
"Don't be afraid, Bryce. It's just a spider," his dad
whispered, stepping closer. "She's more afraid of you
than you are of her."
The word *she* refers to

○ the spider.

○ Bryce's mother.

○ the bear.

○ the wolf.

Answers are on page 117.

Reading

1. What happens to Bryce in this story?

_____

_____

_____

_____

_____

_____

_____

_____

_____

_____

2. Why does Bryce smile at the end of the story?

_____

_____

_____

_____

_____

_____

3. Do you think Bryce is really afraid of bears?
Support your answer with details from the story
and your own ideas.

_____

_____

_____

_____

_____

_____

Answers are on page 117.

Writing

4. Describe something you are afraid of. Explain why. Remember to check your spelling, grammar, and punctuation.

_____

_____

_____

_____

_____

_____

_____

_____

_____

_____

_____

_____

_____

_____

5. What do you think you would like about the adventure Bryce had? What wouldn't you like?

_____

_____

_____

_____

_____

_____

_____

_____

_____

_____

_____

_____

_____

Answers are on page 117.

Answers

The answers for open-response questions are **sample answers only**. The EQAO expectation for an open-response question is that it will

- address all aspects of the question fully.
- use specific and relevant details, information, or ideas from the selection.
- communicate ideas clearly.
- use the conventions of spelling, punctuation, and grammar correctly.

The Best Worst Birthday Ever

Multiple Choice

1. a short story
2. It's too big.
3. everyone had fun.
4. the cake wasn't chocolate
5. cranky
6. everything
7. Where's the cake
8. Finally, all my friends went home!

Open Response

1. It was a disaster because everything went wrong. There was too much going on, and the home was too small for a circus.
2. It was a blast because it was a lot of fun and there was a lot to do.
3. I would like to have a circus party with lots of animals, but I wouldn't like the animals to be running around everywhere. I don't think I would bother with the circus tent, but I would love to have lots of clowns.
4. The main character, Elaine, is a young girl who doesn't seem bothered by the odd things happening around her. She has lots of friends and likes her parties to be interesting.

5. If she invites a zoo to her next party, as the story suggests, I think it will be even crazier than having a circus. There will be lots of animals running around. Maybe her guests will get hurt. How will she keep the lions from eating everyone? The monkeys might throw the food around or eat it all. The elephants and giraffes won't even fit in the apartment. It might be very stinky too, because the animals are not house-trained. And she would need a lot of trainers to help keep the animals and guests safe. It would be a very expensive party.

The Adventures of Two Mice

Multiple Choice

1. in a city
2. All of the above.
3. the cows
4. grain.
5. tasty
6. bicycle
7. Isn't it great
8. How will we get across the street?

Open Response

1. I think the author chose this title because both mice are going on an adventure. When you go on an adventure, everything is new, exciting, and sometimes surprising.
2. Howie didn't like that there were so many people in the city, and he was scared he was going to get stepped on.
3. Mort is a city mouse who likes to be where it is busy and there are things going on all the time. He's a white mouse with big ears and a long pink tail. The sunglasses and vest he wears make him look cool and casual. He's a mouse who knows his way around the city. He knows what to do and where to go. In the country, he seems sort of lost and frightened.

4. Answers will vary. Check that the choice of place to live is clearly articulated and supported with reasons or details.

5. Answers will vary. Check that the journal entry explicitly and clearly describes a place and what one might do there.

Force Detective

Multiple Choice

1. quickly.
2. pulls Pudgy back down to the floor.
3. to remind her father to buy milk.
4. another name for a private detective.
5. hungry
6. there
7. "Dad! I'm home!" I shouted.
8. suddenly

Open Response

1. Pudgy demonstrates to Flora how gravity is a pulling force by knocking the cracker off the plate and by jumping up in the air. Gravity pulls him back down.

2. Flora rolls her eyes at her father because she doesn't think his joke is funny.

3. Answers will vary. Check that at least two healthy snacks are listed, such as apples, bananas, cheese, carrots, crackers.

4. Healthy snacks are important to give me energy, vitamins, and minerals, without too much salt, sugar, or fats.

5. 1. Get out two slices of bread. 2. Spread on jam and peanut butter. 3. Put the two pieces of bread together. 4. Enjoy!

Saving Sand Dunes

Multiple Choice

1. some of the sand gets pulled into the water.
2. breaking the rocks down into sand.

3. Branches stop the wind from pushing sand off the dunes.
4. Plant roots help keep the sand in place.
5. damage
6. Slowly, a dune starts to form.
7. their
8. walkways

Open Response

1. Sand dunes are formed from the pulling force of the waves and the wind. The moving water from the waves erodes rocks into small pieces of sand. The waves push this sand up onto the beach. The sand on the beach begins to form a small dune. The wind pushes more sand onto the dune, making the dune larger. After some time, plants and grass may grow on the dune.

2. "Saving Sand Dunes" is a nonfiction piece of writing. Nonfiction writings are true and usually full of facts. This story includes true information that teaches the reader about sand dunes. Many nonfiction texts include sections with new information that have their own titles. For example, there is a section titled "How Can People Protect Sand Dunes?" in this selection. Also, many nonfiction texts include tables or photos that explain important information. An example is the photo of the fox hunting for food.

3. I helped the environment by organizing a community clean-up effort, which included picking up garbage in our area and throwing it out, and recycling items such as water bottles and discarded paper.

4. Wow, was I excited when Mom told me we were going to Hutchison Beach this morning! We went early in the morning so we could get a good shady spot under a tree. Mom brought a picnic basket full of sandwiches and healthy snacks, as well as a lot of water to keep us from getting thirsty in the hot sun. We went swimming in the lake. The water

was cold at first, but once I got used to it, it was very beautiful and refreshing. After spending almost five hours in the water, we decided to go home. I'll never forget this wonderful day.

At the Sugar Bush

Multiple Choice

1. late winter
2. a sap pan.
3. The cloth is used to filter dirt and ash out of the syrup.
4. the sap
5. pour out
6. while
7. It's going to be a perfect day to make maple syrup.
8. Abbey

Open Response

1. First, you hammer a metal tube into a tree. You hang a pail under the tube. Sap drips out of the tree, through the tube, and into the pail. You collect all the sap and pour it into a larger container. You boil the sap until you get syrup.
2. I know that she likes being in the sugar bush because she talks about it being a perfect day and that she *gets* to pull the sled.
3. Answers will vary. Check that there is a clear viewpoint expressed and that it is supported with details.
4. Answers will vary. Check that the letter includes a salutation, the date, a clear communication for the destination supported with reasons, and a sign-off.

Jane Asks Her Grandfather

Multiple Choice

1. She was doing a project for school.
2. walk over deep snow.

3. She thought metal was needed to make tools.
4. He was pleased that Jane was learning new things.
5. cried
6. The first thing I wanted to know was what foods our people ate.
7. because
8. happy

Open Response

1. Jane learns that her nation used to make sharp tools from stone, wood, and bone. Today, most tools are made from metal. Today, people mostly use cars to travel from place to place, but long ago members of her nation used canoes or snowshoes to travel.
2. They made their canoes from birchbark because the bark was light, and this made the canoes easy to carry.
3. Elders are very important because they are wise and good at listening. Elders help people in the community by listening to their problems and offering them good advice. Elders are also teachers, and they explain history and traditions to younger people.
4. Answers will vary. Check that there is a clear comparison drawn between childhood now and then.

A Letter Home

Multiple Choice

1. in Scotland.
2. keeping the woodbox filled and getting buckets of water.
3. a group of people working together on a project.
4. They dug potatoes.
5. gather
6. because
7. We had a logging bee, so we were able to build a snug log cabin.
8. weaver

Open Response

1. Jamie's mother said, "We're all toughening up." because they are working on the land and the work is difficult. Everyone has to do many tough chores, for example, churning butter, carrying sacks of wheat, digging potatoes, and gathering water, nuts, and wood.

2. I think Jamie wrote a letter to his cousin because he wanted to describe what life was like in Upper Canada. Jamie also wrote the letter because he missed his family in Scotland and he wanted to know if his cousin was going to move to Canada.

3. I would like playing with all the animals on the farm. I would like helping my neighbours and getting to know them.

4. I would dislike all the hard work that I would have to do and being in a new country far from my relatives.

Belling the Cat

Multiple Choice

1. a tale with a moral.
2. the cat lived in the barn.
3. She knows no mouse will be brave enough to bell the cat.
4. The bell will signal when the cat is nearby.
5. grumbled
6. "You're right!" agreed the youngest mouse.
7. After all, it is a cat's job to chase, catch, and then eat mice.
8. frighten

Open Response

1. The moral of this story is "Talk is easier than action." All the mice agreed that putting a bell on the cat would solve the problem, but the grandmother mouse knew that none of the mice would actually do it. It is easier to come up with a plan and talk about it than it is to carry out the plan.

2. The farmer moved the cat from the barn into the house because there weren't any more rats in the barn for the cat to chase, catch, and eat. The farmer may also have moved the cat because the farmer wanted to get rid of the mice in the house.

3. Answers will vary. For example, the story might start in the following way: It's nice to come into the cozy house after chasing rats around the barn all the time, but now I have to get the clever mice out.

Wiggly Worms

Multiple Choice

1. bodies made of segments.
2. small hairs.
3. Its skin dries up.
4. why the worm is tired.
5. scare
6. It's a lot of work twisting, tunnelling, and eating through the soil.
7. under
8. earthworm

Open Response

1. Worms eat soil, dead leaves, and insects. They eat their weight in soil every day. A worm uses its mouth to eat the food. Worms do not have teeth and use their muscles to digest food. The food slides through the worm's body into the crop and down to its gizzard. After a worm finishes eating, castings come out of its tail.

2. Worms are helpful to plants because they turn rotting leaves and animals into food for plants. They do this by creating castings, which are full of nutrients and minerals. These castings are the food for plants and help them grow. Castings also help soak up water that the soil needs. Also, when worms slither underground they mix the layers of soil together, and this also helps plants grow.

3. Answers will vary. Check that ideas about what to include in the garden are clearly communicated.

My Desert Home

Multiple Choice

1. a hole

2. to keep cool.

3. a proper noun.

4. It explains how desert animals survive in the heat.

5. barely

6. their

7. Most plants that grow in the desert have long roots.

8. For

Open Response

1. Ada says "Wow, are they wrong!" because she knows that there is much more to the desert than sand. For example, there are many types of plants and animals in the desert, and there are oases with underground rivers.

2. Most of the plants that grow in the desert have long roots that reach water deep underground. Some of the plants store water in their stems. There are also plants that will begin to grow only when it rains. Most desert plants have seeds that can live in the dry soil for years.

3. The Sahara Desert has a lot of dry sand. The sand is very salty because the Sahara was covered by seawater long ago. The desert sand doesn't have a lot of organic matter and isn't rich like other types of soil, but it is full of plants seeds. There are also many animals that live in the sand.

4. I know that Ada is proud of living in the Sahara Desert because she uses an exclamation mark when she says that her people know "all about the desert!" Also, she tells the reader that her desert is beautiful.

Teeny Tiny Tina

Multiple Choice

1. a small, clever woman with a great big voice who loves to help others

2. a teacup

3. someone clumsy.

4. she saves a lot of money that way.

5. It introduces the main character and the setting.

6. them

7. "HEY!" she would holler, "THIS IS MY PLANET, TOO!"

8. because

Open Response

1. Tiny Tina lives in a teeny house. She uses very small objects like we would use large objects, such as an acorn cap instead of an umbrella. She is a radio-show host who makes speeches. She has lots of money that she gives away. She travels a lot, but she can take herself and all her luggage in other people's pockets.

2. The writer describes Tiny Tina as "terribly blunt" because she tells other people exactly what she thinks about them. She even calls other people names like "BIG OAF."

3. Yes, I think I would like Tiny Tina because she's smart and generous.

    No, I don't think I would like her because she's too blunt and loud.

4. I would like being so tiny that I could travel anywhere in another person's pocket. I would like to be able to dive or swim in a glass of water. I could crawl through pipes. One chocolate bar would probably last a whole year.

5. I don't like Tina. She's rude and mean. She's always yelling at me. I might have almost stepped on her once or twice or three times. But that's no reason to yell at me. She can be really kind and generous, but she can also be really mean. It's not my fault I can't see her.

6. Answers will vary. Check that a salutation, date, and sign-off are included.

"I'm not afraid …"

Multiple Choice

1. Spiders scare him.
2. a fear of wild animals
3. hiking and camping in the mountains.
4. Bryce likes to take photos.
5. soothingly
6. "I'm *afraid* of falling!"
7. "I'm not afraid of the bear! I was just surprised," Bryce protested.
8. the spider.

Open Response

1. Bryce goes on a hike with his parents. He sees lots of things that scare him. For example, he sees a spider and is startled by a bear. His parents try to support him, but he finds them sort of annoying. For example, his mother is always telling him the technical word for his fear and what the Greek words mean. In the end, he is able to turn things around because they are afraid of the dark and he isn't.

2. Bryce smiles because his parents are the ones who are frightened, and he isn't afraid at all.

3. Yes, I think he is afraid. I would be afraid if I saw a bear so close. No, I think he's just startled. He turns a corner in the path and suddenly sees a bear. That would frighten anyone.

4. Answers will vary. Check that sentences are complete and that the description includes adequate detail.

5. Answers will vary. Check that preferences are clearly worded.

Test Tips

The following tips will help you be successful on the EQAO Literacy Test.

Reading Selections

1. Read each selection carefully. Think about what you are reading. Use reading strategies such as visualizing or making connections.
2. Think about the details in the text. For fiction selections, you might be asked questions about the setting, characters, or plot. For nonfiction selections, you might be asked questions about the main idea or about supporting details. As you read, identify these aspects of a selection.
3. Reread each selection. Focus on any parts of the text that you might not have understood the first time. Underline key details.

Answering Questions

1. Read each question carefully. Think about what it is asking.
2. Ask yourself: Do I know the answer now? Should I skim the selection looking for the answer?
3. Look in the selection for key words from the question.
4. For multiple-choice questions, think about each possible answer. Which answer is the best response to the question? If you're not sure, review the selection. Cross out any answers that you know for sure are not correct.
5. For open-response questions, make sure you understand the question. Skim the selection, looking for details to support your response.
6. If you don't know an answer, move on to the next question. If you have time, return to that question.

COMPLETION CERTIFICATE

CONGRATULATIONS!

You have completed the *Nelson EQAO Grade 3 Reading and Writing Workbook*!

Presented to:

_____

Date:

_____

GREAT JOB!